LETTERS IN THE DARK

HERBERT LOMAS

❋

Letters in the Dark

Oxford   New York

OXFORD UNIVERSITY PRESS

1986

Oxford University Press, Walton Street, Oxford OX2 6DP
Oxford New York Toronto
Delhi Bombay Calcutta Madras Karachi
Kuala Lumpur Singapore Hong Kong Tokyo
Nairobi Dar es Salaam Cape Town
Melbourne Auckland
and associated companies in
Beirut Berlin Ibadan Nicosia

Oxford is a trade mark of Oxford University Press

First published 1986 as an Oxford University Press paperback

British Library Cataloguing in Publication Data
Lomas, Herbert
Letters in the dark.—(Oxford paperbacks)
I. Title
821'.914   PR6062.045
ISBN 0–19–251959–3

Set by Latimer Trend & Company Ltd., Plymouth
Printed in Great Britain by
J. W. Arrowsmith Ltd., Bristol

For Lucy and Matthew

Acknowledgements

Acknowledgements are due to *Ambit, Country Life, Encounter,* the *Honest Ulsterman,* the *Hudson Review* (for I, XIX, XXIV, XXIX, XXX, XXXVII, L), *London Magazine, Outposts,* the *Reaper, 2PLUS2.*

Ma tu, perchè ritorni a tanta noia?
perchè non sali il dilettoso monte
ch' è principio e cagion di tutta gioia?

Inferno, i. 76–8

I

I lay awake writing letters to you
In the dark—so I got up, felt alone,
And put on the light. I need someone.
There's a kind of intimacy that's closer than the bone
And can sometimes get into letters, more than anywhere.
It's communion with the dead, or more like prayer.
The dead are somehow refined, or are being refined.
When they answer back it's in disembodied voices
From which all static and chatter stored in the cells
Have been magnetized off, buried, or even burned.
In the purgatory of the small hours the person
God might have first imagined stirs on the rock,
Pushes the vulture from his liver and begins to turn
His anguish into intimate words, or at least
Disembodied expressions of love disguised
As chatter and sent to a disembodied love.

II

So here I am in my usual pew, between
A supine Shakespeare—crapula must have struck—
And generous John Gower, gaudy as Blackpool Rock,
And stiffened flat, it seems, with atropine;
And, slightly in front, our spastic lady's got a sticker
On her wheel chair: My Other Car's a Porsche.
She can read me: her senses are quicker
Than ours, and she corkscrews at my applause—
Which is entirely in my head.

                The Ghost
Is close to the distressed (I know) and she
Can bless without trying, easily understands.
Tears embarrass my eyes: I have to see
Through wet refractions our filtering blue host
Astride the world like glass, with doves for hands.

III

Outside, the competers: competent and virtuous
(Often ourselves); here—incompleteness; screening glass
Chequered with flesh but letting stained light pass.
Worldliness to the rail—all words superfluous
And inexact, who hardly know what ails us.
That's why we're here. We're usually crass
Or stony with good and evil, yet Christ's Mass
Invites the knower to be dubious.

It's our goodness and our knowledge he can't use:
They make us so benign—put us in a class
Apart. Stiff with restraint, how messy if
We get crushy with love and play the silly ass!
Behind the unreadable fool the unspeakable hieroglyph—
And hermeneutics here would merely confuse.

IV

Insight is sousing in: grizzly mist on the horizon.
My sea's very upset: as if it can't
Decide to ebb or flow—yet the sun's
Quite tense, like a discus—or poised for a sprint
From cloudhole to cloudhole; rack hits salt, and everything
Inside me tastes of brine—and there's so
Surprising much it seeps over in tears, stinging
Gladness, gratitude, unreadiness—sorrow!

Listening carefully to the sea—those huge hugs:
Following its wriggling fingers, eeling down
And beckoning along drifting strata to the tug
Of intense cold deep down below: shown
Something quite different from anything you've seen before—
Steering fishily up from an unknown floor.

V

A week ago, what should I see but two young swans
Who'd found each other; asleep in a ditch; full-grown
Cygnets, brown on a green marsh mattress
Of duckweed; dismantled wind-up gramophones . . .

Their beaks tucked back under a wing
Like fold-up needle-heads—inward-bound,
As self-contained as two pies, till she felt me
Looking and unsheathed her head, peeking around.

Then he woke too—quizzing more carefully,
Solicitously, more protectively than she,
But, seeing no danger, turned a serpent's gaze
And studied her with deathless *agape*.

They considered sleeping again but felt a need
Had come to paddle on; so, with slow care,
Each tested the stretch of a brown wing
As if to know the gear was in repair—

And made two partings down the duckweed,
Not parted in their partings. I had to know
Swans love with inseparable love, and so our birdselves go
Treading the water, not knowing on whom we feed.

VI

Shakespeare buried his actor brother
Here, next to Massinger,
Who neighbours Fletcher,
His honeyed collaborator
Who might have lived longer
But for a septic delay
In plague-ridden London
To visit his tailor.

1607: Edmund Shakespeare:
Just a year after
King Lear.
The sibling joke
About the bastard wicked brother
Might make a brother choke
As he stood where we stand.
Those family names:
Hamnet died like Hamlet.

But Massinger, Fletcher, Shakespeare,
Henslowe and Alleyn,
Men of this parish
Who swashed and buckled for gain,
Have parted from their profits,
Actorish or bearish,
And their names,
Which we cherish,
Their quibbling for King James,
And sprung to that perpetual spring
Where no benumbing cold or scorching heat
Can diminish anything
Of what hangs here for ever
Suspended in this air:
What men endure, the life of prayer.

Famine nor age have any being there.

VII

If the brain's the mind's rind
That we grip through the many senses, and striped wasps
Sting and build as they do from a habit
They use like a pension, which works, though it keeps them
Little machines, with genes for mastermind,

And man's the timeless but evolving kind, the least
Grabbed by habit, copying, creating, hawking
New science and new solutions
That clock and alter his heart and will in time
Draft how his body will officiate and feast,

Not part goat, part ghost, but
An outward and visible cipher of an inward and spiritual
Contriver, why does an old man polish brass
And peer from a door, as if for his life, wondering
Why it never arrived, or trail his foot

Past a funeral parlour that says 'We're here to help you'?

VIII

Yes, you will smile, but does the mind exist,
The spook in this machine? But does the face—
Broadly, that is, as we think it does, in space?
Looking at what we call a brain or fist,
Who is it that perceives? Not that little cyst,
Or camera, the eye: no image or embrace
Pulses up the nervous tracks to race
The dendrites, tell you you've been kissed.

The brain is something we've imagined.
Our ganglia jerking, electricked
Into ecstasy, who crooks this underpass
Of input into insight? The flesh is grass,
An adoring face, or a twitching skin of conflict,
But who concocts these molecules into man?

IX

The hoisted leg and waddle of crows,
As if they'd no business with the grace
Of birds—just a job to do,
Eating, practical,
Claws on the ground—except when they
Lengthen out their finger-ended wings like hawks
And waft slow-flapping off
To another dung-heap.
                         These undertakers
Seem so much closer to
Men than a blackbird.

X

I've forgotten the scruffy intellectual I once was.
Can I admit the divine man in friends,
In the hips of a camp Indian, in fag ends
Of toothless tramps, washed policewomen, or the cross
In a class snub? It's my loss
That I'm one of the clean who condescends.
I scare the needy—my clothes can stink of dividends.

What God would want to know me? Though I'd like
To praise him, I stare out of a bay window
At a chirping garden, a holly tree, conker spikes,
A knowingness of late roses. They could imply
A mask of sorts: the passivity of this show
Can blink like a face—everything in it an eye.

XI

The pelican hacks her breast and suckles with blood:
Guilty pap of fledgelings of the good
Who badly devour the edible breast—the musky silk
That circumvents with luxury our milk.
Wagner touched silk and cunt—smell and love of skin
That dies with love of death. The nectarine
And curious peach call up the ticklish nipple, even when
Men prefer gardens. Yeats's famous interest
In death was but his longing for the breast
In Eden gone: he pressed and was impressed.
And how to let him out—the child inside
Who knows no time, our nosy cause and guide:
Love would distend him: he knows what's best.
Babies nurtured at their mother's breast
Dream in unison with her, as if in the womb,
Even when they sleep in a separate room.

XII

Innate science:
A martin inherits a technology
Of clay pellets: a genetic testament.
Kaspars don't build houses with their genes.
People learn what they can, repeat and imitate,
If pushed invent.

Words we have—almost like causes.
But a new elusiveness smiles
From our new names
Which give us power to shake
The land that's barren now and cursed
For nomenclature's sake.

XIII

Lionel Lockyer claimed his pills
Had been distilled
From the rays of the sun:
They'd keep the agued carrying on
In fogs and the contagious air.

On his memorial bier
He seems to have a migraine
And a bicentenary grin
As one by one year after year
The tourists peer

At his well-wrought epitaph
And laugh.

XIV

There's something we're always missing among
These antiquated people. They're strong
On only baring themselves to their personal sex.

They finger their thoughts by themselves,
With whisky bottles behind bookshelves,
And phantom touch, no fondling to perplex.

What could be fairer than their class
Dues, but the upper upper are crass.
All kin are not brothers, or lovers.

At odds with bodies, they're cowed
By actual pricks and breasts, though proud
At mirrors, quick to see spots on others.

Find them attractive, and give them a hug, and they
Rise to you, but quick—they pull away,
To bespatter you later out of rectitude.

Yearning for those shudders and delights,
They have to cope them in accepted rites
That prettify a calculating attitude.

One perfect intimacy: nothing else will do.
But it doesn't. The lies prick through,
And sweet imperfect loving is subdued.

XV

Ave verum corpus: we're here—and halt—
Because we're feebler than others, rock salt.
We limp with sclerosis, a stick for senile thighs,
An arm to help these lightweight brittle bones to rise—
And each audacious step a precipice
Of hospitalization yet again: indignities—
Flat feet, arthritic shuffle, sheer
Ugliness, incontinence, smelliness or beer.

The first Adam was an advanced primate.
We creaking hairless gorillas malinger
To be cuddled, old dogs exposing our privates,
Pressing our itching ears to the caressing fingers.
Yet each has been called to be something else:
He calls us out from ourselves like funeral bells.

And the golden face looks out from the icon and glows,
Supporting all this dole with God only knows.

XVI

The lake crazes into splinters, jaggings, chips:
A wind like an artifice orchestrates the trees,
Screams quietly, crescendoes to a wheeze,
Comes cracking down with thunder, lightning blips
And rain: a tremendous tutti, then diminuendo slips
Squalling away. For ghostlier ears than these
A wind of no coming has knocked men to their knees,
Rocking their standing with apocalypse.

My boat stops on a glassy lake, expects.
Light shines inside me vertically, thrown
Through my head down to the centre of the earth:
A husband comes inside through love, not worth—
Face witnessing face and subject subject.
Amongst a world of glass who knows we're known?

XVII

Hagar's in prayer—but breaks when the angel comes:
She didn't really expect it—or the sudden landscape
Beyond the carnally twisting trees: how to escape
The non-molecular rock, the thumbs
On her ears, forcing her to look
At the river she knows she's got to cross,
That motionless boat, the fishers, the final fosse,
And the bridge weathering to original rock.

It crashes on her head—the great dismay:
No rules, no way,
No place to hide, and this that's just begun
Now solider than the soil she's clung to. Pregnant today
But not with what she cannot bear, she spawns
In the desert a wild ass of a son.

XVIII

Oaks: nervous systems
Against a December sky.
They tickle my dendrites
As my axons try
To escape infinity
For when I die
I'll feel my vacuity
Stretching like sky
And the tree of my body
Left to lie
As I ask emptiness why
I lost identity
Unless it was to try
As I do now
To rediscover entity
After pitching myself away
And learning to deny
In my hard way.

XIX

This fly was born in this attic
With no food. What can he do but fuss,
Plod up a pane of glass, fall,
Buzz, plod up again like Sisyphus.
He was here yesterday, and the day before,
Now he's got to the mullion, crawls
Along it, round the corner, excelsior—
On a higher pane above. Outside, climatic—
All that the glass is blocking him from meeting.
I open the window. Retreating
Upwards from the space above the sill,
He needs some coaxing with a pencil,
Then side-slips into coldness that will kill.

XX

So here's the alarming silly tale again:
Our father Abraham's call to murder a son,
Hearing a voice—like the one near a Bradford grave,
Advising the Ripper to dekink casual love
With hammer, axe and sharpened tool,
To gouge and bash those girls just out of school.

And the father's needed sacrifice,
Provision for a life, has had its price,
Provision for a death. And for the son—
Priapic expensive manhood's just begun.
So boys must learn of life on burning wickets:
Rams caught by horns on convenient thickets.

Yet Abraham's caught by the horn—and aligns on the lie:
Inwit, the great I Am, insights his eye.
We Israelites, the wandering heirs
Of a bewildered dervish, are taught distrust in prayer.

XXI

Brown in the autumn, grass is blown
And brittle, rustling dry: children's hair
Whose eyes have intimated, not quite known despair.
Or wind snows—swerving, like this whistling son
Who's whistling by the church and the shuddering trees, down
By the slugsmeared arch, the start of some new scheme:
The viaduct drips and echoes on its stream,
Each drip a print: you can't guess where, then gone.

Alone, alone—a stony watered echo
Boomed by the sleety archway deep down here.
A train shrieks overhead: and must he go
Hooting with pleasure, trailing fortissimo,
Diminuendo, piano, to cold fear?
The lean wind grips him, twisting and pulling his hair.

XXII

St Cuthbert's hopping—a cassock on the shore—
At two ravens, rooting with their beaks
In the thatch he's heaved in the North Wind's roar
To top the hospice on the lonely island of Farne.
He can't believe their cheek.

'That's a *nest* you've got in your craw',
He shouts. But the ravens just scoff. 'In the name of Christ,'
He cries; 'it's for my friend St Herbert! Yes, caw—
Come here again and you'll get my boot, not straw!'
He's never been like this. They fly off fast.

Yet one comes back regardless. Cuthbert, bending to sow,
Studies the trailing wings, reads what the bird will say.
Beak at his feet, his croak is slow,
Screwing his neck at Cuthbert, and starting to go.
But Cuthbert catches his eye and grins: 'Well . . . stay.'

They both fly back, lugging a present of lard—
A gift you could grease your boots with. And by and by
He lets them live in his yard,
Where they look askance at their saint—and very hard
At the hole he's built in the roof to watch the sky.

XXIII

All my most intimate talkers—they've been
Men and women of inwit I never met:
Survivors, inhalers, whom the threat
Of the matter that matters or chore of citizen
Compelled less than a whiff of ghostly oxygen
Their nostrils flared at: Lawrence, Rilke, Herbert,
Wordsworth, Blake, Hopkins, Bronte, Dickinson:
Abundant life's seductive martinets.

It's no back-suction to the womb:
Rather the opposite. It's the whack
Of the little beak that pecks the shell. Whiteness
Glows through the crackable walls of the once big room
That's shrunk to a bag: one peck
And a great slit opens onto brightness.

XXIV

Now here's a cosy photo:
Two seekers who look quite fun—
Both of them holding roses
And with curious costume on.
One is a mystical lama,
The other a lunatic don.
Both are bedragoned with visions
From *The Tibetan Book of the Dead*,
Though the whole thing is loaded with dread
And could have been once all a con.
But something's gone right with the lama,
And something's gone wrong with the don.

XXV

William Austin looked on his mother's face
And thought: As the clear light is upon the holy
Candlesticks, so is the beauty of her face
In a ripe age. William was melancholy,
Unsure about death, though he joyed in a well-writ book,
A picture and a song. No beautiful thing
Is made by chance, he thought, but even of learning
As of wine, a man may go on hungering
Till it make him mad. Shall there be nothing
Left me but a grave? Shall I at last
No other dwelling have? But till we come
To Thomas, and his confession, we have no Christ.

XXVI

All thoughts are parrots, pretty mocking birds.
Twitter about God's a cultural activity.
Does it show more than intertextuality
To shuffle packs of metaphysical absurds
Or crap out concepts like a Luther's turds?
Reading the Bible's normal creativity:
The reader is the writer: each subjectivity
Fabricates a Jesus from a text of words.

Your God's too human: those signifiers
Were organized by stone-age Jeremiahs,
Linguistically ingenuous desert bards . . .
Which makes the Word a broadcast out of silence.
Intruded in the syntax, truth interlards
The structuring brain's concocted violence.

XXVII

To be sponged on
And neglected,
To have people take
And give cheek back
Is, if you catch the joke,
To be Godlike and grin sheepishly
As you sense the shepherd coming close.

It's in these silly
Moments of blissful dying
That, elect with God
In your neglect,
You recognize the Host
And swallow without trying.

XXVIII

'O God,' prayed Andrewes in secret every night,
'Save me from making a God of the King.' To observe
The grass, herbs, corn, trees, cattle,
Earth, waters, heavens, to contemplate
Their orders, qualities, virtues, uses was
Ever to him the greatest mirth, content
And recreation—held to his dying day.

Shakespeare had little Latin and less Greek,
While Lancelot had a lot of both, as well as
Hebrew, Chaldee, Syriack, Arabick, Aramaick.
'Your Majesty's bishop is a learned man
But he lacks unction; he rather plays with his text
Than preaches it. Andrewes can pray as no
Other man can pray, but he cannot preach.'

Yet if there were two saints of God in England
That summer, they were surely to be found
Under the roof of the Bishop's Palace at Ely,
Alone with their books, and still at work with God—
Abandoned souls—prostrate in the scholar's litany.
Casaubon and Lancelot Andrewes doubted they were
True scholars that came to speak to a man before noon.

'On the Real Presence we agree with the Romans;
Our controversy is as to the mode of it.
We define nothing anxiously nor rashly.
There is a real change in the elements. But the death
On the Cross is absolute: all else is relative.
Christ is a sacrifice—so, to be slain.
A propitiatory sacrifice—so, to be eaten.'

But Privy Counsellor to the royal pirate, fattening
His favourites? To dine at the court of flattery?
To be carried about, feasted and amused
By the heathen poor, who had to find for James,
Laud and Andrewes, horse, carriage, royal board
Or feel the wrench of a rope around their neck?
And your Devotions slubbered and watered with penitent tears?

His Lordship kept Christmas all the year.
'Come and transfix a buck with me. I am
Not well in my solitude: the hand that writes
These lines is ill with ague. Let me see
Your dear face. If Stourbridge Fair, the finest
In England doesn't suit, I have beside me
A Matthew in Hebrew to make your mouth water.'

'A more servile and short-sighted body of men
Than the bench of bishops under James the First never
Set a royal house on the road to ruin.'
Lord, I repent: help thou mine impenitence.
Who am I, or what is my father's house,
That thou shouldest look upon such a
Dead dog as I am. I spit in my own face.

But who cursed the hanged and dismembered Gowries
Every August before King James? At least
One commandment must have exceptions, as we're
The first to admit: our public executions
Are marvels of technology in the Middle East
And help the Exchequer; the saintliest then
Admitted the King's enemies should be hanged.

And so if persons nowadays feel squeamish
Even about hanging, here at home, to say
Nothing of drawing and quartering, it tells more
About the fashion of the age than about
An individual's virtue or conscience. Saintliness,
After all, is vicarious redemption, inviting grace,
Not a capacity to see beyond your own time.

Andrewes prayed with dust on his head, a rope
Around his neck: in great calamity we exercise
Great devotion. A knighthood might be had
For sixty pounds, but only nothing will buy off
The irrefragable accusation of our lives.
Kings' benevolences are malevolences indeed,
But God hath not turned his mercy from our side.

XXIX

By the standards of Parliament
Jesus was very naive.
It's not that we're not all agreed.
We know what's rational,
It's that rationality's not compatible
With powerful interests and greed.

The world's very old,
Much older than the Messiah.
And more modern.
The way things tread and science
Between them
Have trodden and trodden and trodden.

He didn't know
How the world wagged,
The politics of size.
To realize his ambitions
Or even survive
He needed to compromise.

If he succeeded
In ruling the world
It'd have to be thanks
To the Roman Empire
The Sanhedrin
Or nowadays the banks.

Unfortunately for him
He'd thought all this out
And got it wrong
In hallucinatory argument
With the devil,
Who loves the strong.

XXX

On the cover of *Honest to God* there's a depressed
Thinker. Perhaps all thinking is depressed.
It's depression that makes us think. Is it because
Our faith's too cheerful that we miss sometimes the wit
Only those who know the truth's always
Absconding can hit? The Bishop's face
Is blithesome and halcyon. To have faith in God,
Even if 'God', perhaps, is not a word
We should use for a generation, is faith
In more than truth, for Jesting Pilate was inspired
To ask the question of the Way, the Truth and the Life.
The joke was on him, we presume. A reluctant revolution
Is dying daily: difficult to admit
That what we know to be true is no longer true.
As we stand on our dung-heap of ideas and crow,
He slips away whispering 'I Am', to break bread
Just when we're discussing his non-existence,
As we know from time to time but not in time.

And Bishop John Robinson left his scholarship,
Waiting to see what he would learn from cancer.

XXXI

Elizabeth Newcomen once sold milk,
And Dorothy Applebee married a brewer.
The money in their legacies for beggars
And the schools they built
Serialize immaterially
In children not yet born
And those who don't know where their luck comes from:
Even goldener than the gilt
Of Dorothy's candelabrum
Still dangling like a rumour from the tower.

XXXII

The shark survives
Although he doesn't chatter
Like whales or major in science.
The quality of survival
Doesn't matter.

We're pickled in our own solution.
The murders we watch on the box
We think we have to do.
History's watching beside us.
The skeleton that knocks
In the pipes is fossiled here inside us.

Is there a leap we're still not trying?
The dog that groans in travail on the floor
Is waiting for me,
Whimpering and sighing,
And feeling unspeakably more
Not less than I,
As he twitches with loving fear.

I'm always saying goodbye.
He's always here.

XXXIII

And everyone who's bitten to the core
And cannot taste the apple—
Can they stare in new confusion at our pain?
It's not enough to feel the ripple
Of the twisted thigh within our brain—
Who can see more?

Who can get at what's so absent,
That casts the longest shadow,
The sheer black dazzler of our nature,
The blazing lover of Teresa,
The shining dark within the brain
That Coleridge found and, finding, foundered on,
That Blake sang songs to and Wordsworth saw and mourned?

Shall we who timidly built up our seclusion
Split wide the walling skies like troglodytes
And clamber from our cavelight and amnesia?
Can we see the whole race walking, hair now bright,
Through shocks of heavenly seizure
In the glades of beautiful light?

XXXIV

We look for chocolate wrapped in silver paper
Or Father Christmas, in from the planet Venus.
But he lives in sin and points to wounded feet
And hobbles round in distant labour camps
Or dies unnoticed in a neighbouring street.
He whispers kindly when we start to eat him
And peeps at us from eyes of smelly tramps.

And, peeping back, we feel we're smelling death
And turn discreetly from distracting life.
His scripture is the dirt we're often doing.
He points our noses at our petty strife,
Demands the loaves and fishes from our pockets—
Provisions we're withholding from the war—
But adds, to those who never feel accepted,

'Accept that you're accepted—as you are.'

XXXV

On the summit of St Thomas's Hospital tower
I felt no doubt. Easy to imagine
The buses and formicating cars
Queuing for regroupment over Lambeth Bridge
Were the last glints of a world I was leaving.

The Thames winked, and as the optical day
Staged its illusory matinées of light,
The incredible stained to credibility, like
God's love at a canine remove, transferred
From the pack to man, dogging with joy and healing.

Infants who know every passion before they've
Felt the events they won't understand
When they happen have the unwrinkled skin
Of those who know the story before they thrust
Their hands into the slits of the bloody wounds.

XXXVI

We meet again at the kneeling place
We plod to without emphasis,
Where it's no disgrace
To face another face
As psychic invalids
With lowered eyelids.

In every lowered glance
We find provision
For the journey we're on.
Grace whispers grace
To his own face
Through all these faces.

XXXVII

Midnight Mass: Mervyn tells us
Divinity desires a world like this burning tree.
The choir carols from every coign in turn
Of the semi-darkened sanctuary,
Tuning the stones
And topping them up with that plangent wine
The ghost mouth tastes in every graceful shrine.

I read the prayers from the lectern
And from all those bowed heads
An almost tangible breath has begun to flow:
My heart turns red
In the expanding now
And vertical as a chimney
I feel their flames go through me to the tower.

XXXVIII

Listening's dissolving. After I go—
Surprise: I'm still here. I grasp for myself
And the ungraspable holds my self. I know
Him too, but I eludes me. I am, identified,
Is the person I've just been, an ego in my head.

I am in endless space alone, aware
Of watching as I'm watched, in a bugged cell
With a one-way mirror, which is everywhere.
I watch a bright screen in the dark, but if the light fails
The emptiness reminds me I am here.

XXXIX

Dear God, though I'm sweating round the park for death,
It's not because I'm not with every bronchial breath
Running for life: slow turnings make the book
Seem seamless. It's only when I've lost the track
In comminating thunder, thrashing through flashes,
In crashing streets, with no house, friend, hope of sanity
Or end, no love, no faith, no charity,
That I see the faces I've been gutted to ashes.

Yet deaths like these are not what make me say
Death makes new tracks. The black
Doors grieve open and, yes, I see the Easter Day
Breaking like spring—outrunning dark,
Because you come to meet us in your Mass
And make us see through wine as if through glass.

XL

Whisky's a risky aspect of the Body:
A distilled divine. All sacraments are dangerous.
Ecstasy can booze you to the Devil—
The Dog of God: *Deus est diabolus*
Inversus: the Hounds both hunt the world
As a team (see *Job*) and often it's the Dog who pads
With the prey to God. Whisky killed my dear
Old dad and that's what drove my mother mad.
Death's certainly the alcoholic's mother:
Those first erotic bubblings at the bottle
Or breast are gaspings of our love of Her
Or Him or Both, all Three, met shuddering as we rattle.
How long will it be before we find Cockayne
Where the whisky rivers flow, so stilly flow,
Like every poison from the flesh of God,
With peaty water, heart of gold, an amber glow?

XLI

Imbecile to the wise and sane, our baby divination
Is milky eyes staring at bosoms to be hugged
Not clouded in concepts: an infinite sky's plugged
Into an infant's irises of contemplation.
Impatience
At lugging Moses' stones is right:
The yoke must be easy, the burden light.

The lost are often first.
Sprinting from God
They won't be coerced.
They see him at the winning post,
Give way to a crush,
And fall like fools while the prudent only plod.

XLII

I hardly ever meet my brothers and sisters.
Here perhaps on holy ground
Is the proper place,
But what we pass around
Is the shining handshake, called the kiss of peace.

What can I do but intercede for the publicans
And drinkers, not here, but my actual friends?
There must be, I suppose, quite special roles
For every Madeleine. Who knows what He intends?
And is the kiss impetuous in cautious souls?

My cronies are didactic: one showed me
A film of far more breathless peaceful kisses:
'These rather lovely girls and ugly chaps'
Had overcome the prejudices
'That make us buy these films and watch perhaps.'

They seem to be quite enjoying what they're paid for,
So prodigal and naughty with caresses;
But surely it's the Producer
Who asks for tender smiles and licking kisses
And makes each woman long for a seducer?

XLIII

'Bore men if you must,
But must you bore God too?'
Demands Isaiah:
Mervyn's maiden text
To fluff the dust,
Knowing it might backfire.

Or take a houseless pair
Out of the fifty thousand,
And their baby and an ass,
To construct a living crib
In Trafalgar Square.
The bods on the beat don't like it,
Nor do the brass.

'We've got a crib already—
Next to the tree . . . and hark!
The carols are starting up
On St Martin's steps.'
Mervyn and Trevor plod
With mitre, crook and nod
To a legal plot in the park,
A nook where no one'll see
Their demythologizing cot but God.

XLIV

Of hem, that writen us to-fore
John Gower is erst in loves lore.
Tho he be dede and elles were
We thenken on his presens her,
Tho other worde of hem nought is,
Aperte fro thet he writen has,
Then mariage lisens and his wille
And tomb that stondeth to us stille.
A squere he was and born in Wales
And wened it gode writen in tales
Somewhat of lust, somewhat of lore,
Tho loves lust and lockes hore
In chambre accorden never more,
And leved he in time of blody kinges
And civil warres and tresons swinges.
But Gowerlond, tho fond in Wales,
More sothely is a lond of tales.
Ther ben also that seyn he was
In Yorkshire born from Conquerors,
And otheres seyn in argument
That he was born and bred in Kent,
But as a Suffolk man minself,
Tho adopted so, boghten in by pelf,
I wene he was a Suffolk man,
That leved in Multon manor a span,
And atte leste hadde frends in Kent
And eke thereto establishment.
Dan Chaucer was his poete frend,
A yonge prentice, of hem ytrened.
He preies for Charite, Merce and Pite—
Kindenesse herd to finde in a cite—
His hair hath roses intertwined
With ivy—tribute to his minde
Versed both in bokes and science,
Of which he made an holy alliance.

But Goddes curse had yfallen on
His lond in gret confusioun.
He wended out to gathre flowres—
Insted he sawe a lond of hores
Chaunging into forme of bestes:
Swine in the cite atte theyr festes,
Asses weninge they weren horses,
Dogges that licked the children's sauces,
Oxen proude of dragones tailes,
Venus lost in a lond of wailes,
And so his *Liber Amoris* sholde be
On how men moten love joiesle.
Swanne at his necke, at his fete a lion,
His sawle hath victore oer angeres scion.
Preie for his sawle, and as your guerdon
He boghte thusand dayes of perdon.

XLV

My chest crackles with bronchitis—
And I ask myself, What is it then in me
That's so unhappy with the air?
Air: it goes to my blood, it's everywhere,
It's deeply intimate with my terminal skin.
Cloud teashops, lakes of anodyne,
They float up there,
And here
Her breast and her vagina—
The air's been teaching how to breathe them in.
My lungs are touch, the kiss of my eye still finer.

Enclosed in stone, I breathe through stone like skin.
The cathedral's ribs and groins
Can splice me to all space—
Another air and psychic terrain
Extended by the cinematic brain
Sustained by grace.

XLVI

When I was thirty-two I got cancer:
Probably a proxy for something else,
But it took me by surprise. Time was unhurried
In the specialist's ante-room. To me it seemed
Clouded with witnessing comforters. I imagined nuns
Spilling their love in prayer for the sick and worried.

The specialist ogled it through a tiny lens
Like one I'd had at school. He didn't think
It was cancer. Better to have it off, though. No
Innuendo—but doubtless truth in that.
Two weeks later I'm taut on a clinical table,
Waiting for stitches out and ready to go.

A houseman came in first in a pure white coat,
With high heels, and leaned against a wall,
Hands folded behind his bum. His smoky eyes
Were observant I knew to see how I would take it.
The specialist bustled in, with a hand-wag,
Assuringly: 'It was cancer.' No surprise.

He'd examined every cell. There was healthy flesh
All the way round. The nurse was looking embarrassed.
Who's not heard of people cured of the dread
Disease, and only a year later—they're dead?
Melanoma: a beautiful name for a fast,
Distorting and painful killer when it spreads.

So I needed a little treatment: 'belt and braces'.
Regular visits for radium: massacre the cells
In case a wild one escaped. And so I rubbed
Shoulders with the dead and dying. I remember a lady
Carefully lipsticking lips before a mirror.
My reprieve could make me somehow feel quite snubbed.

The scar will always be mine: it makes me nervous
With women: not gold, but the size of a silver guinea—
And a wound in the side—but the other side from Christ's.
It'd be rather fun to lie: a piece of shrapnel
Got me here, instead of my cushy war.
It looks like a brand on a pig about to be priced.

It only exists in my mind. I conned my surgeon
To carve it out—merely to leave a long
White scalpel line like a sabre wound.
On the beach it won't raise questions: more discreet.
I remember a friend in a pub and remember thinking:
'He'll be kissing the girls when I'm long underground.'

Dead now. I'm here but in his club.

XLVII

Somewhere there's a man who took me fishing
And kept me sane. I don't even
Remember his name. Stephen . . .
Some saint's name: unintellectual, observant, well-wishing.

This useless work has risks. Near to splitting,
Rilke thought his grit in loneliness came
From glassy evenings on Capri, all the same,
With two old dears, in a deck chair, watching them knitting.

One of them sometimes had an apple to pass.
A tiny ugly dog with sorely swelling boobs
Craved for his eyes in its solitude. The sugar cube
He gave her was a wafer in a private Mass.

XLVIII

Perhaps those childish ecstasies of terror
When midnight pulsed on midnights of blind hate
And vampires hung behind the stirring curtains
Alert for children's eyes to close and sleep
Were deaths we might have had or can anticipate.
How we need to have an expert mother
To sit beside us, spreading unconcern,
And mock the terminal illness in the corridor
And tell us that it's not our turn:
The custom-built disease has gone next door.

Perhaps in every aged impotent craving
The child we were is still inside us, crying
For a rocking knee, and not believing
What we heard can never be,
Though finding mother's presence near the sea:
To feel the lift from fifty fathoms
And know annihilating pleasure
Of sea absorbing self in sea, an awe
Far greater than the logic
Of non-acceptance on the certain shore.

XLIX

Some day I'm going to have to meet my mother
Again. I'm not sure how it will be.
Will she be splashing that arctic fox she wore
For the sepia studio plate they had of us three
A month or two after my birth? I must feel kin
With those still girlish eyes and lenient lips—
Pleased, my father proud. I'd find her smooth skin
And neck attractive, her kind intelligent hips
Seductive, her Parisian costume and pearls, and her
Silky hat all that could make a baby purr.

I can hardly pray for her soul. I feel meagre
Because she died before she died,
Too soon for me, and recouped as one too eager
And helpless to help, in need herself. I sighed
For an intellectual love, all nous
Without the feeling. I gave her the best books
To read; she read them greedily. It was
Too late. The lesion had been scored. The crux
Was now: survival meant to disregard
The conscience she gave and find a new mother instead.

November: the month of holy souls. I'd like
To feel gratefully to her for those first baths,
Her cherishing hands on my skin, those grey eyes
So tender even in her stubborn wraths,
My baby safety under it all. But misgiving
Had made her need me more than I needed her,
And what I feared the most was her fear of living.
The day she found us infant boys and girls
Exploring bums in the bathroom was the worst.
She wouldn't speak, and both of us felt cursed.

No, the worst . . . Yet is she listening now? Who else
Could understand her like her only son?
And who could understand me like herself?
She knew me even if her mind went wrong.

She wished me well, though neither of us knew
The way to the other's happiness. Surely now,
Sophisticated in another's life, she can review
More expertly than I. She must know how
To nurse me to myself when I too leave behind
All the crazy blocks that pave my mind.

I'll need some friends who've been there when I go.
As my breath scratches the last scribbles, and I write
'The End', letting go of everything I know,
I'll need a helping hand in those streets of light.
Can anyone love the glaucous eyes of weakness?
Though I've got through life with the back of my head
Turned to most neighbours, trusting in obliqueness,
The Mother will lean one night by the cot of dread
For my sensible eyes to close in sedative calm
And the moment when I perceive who I really am.

L

There's no abiding city here.
A gentleman's
Enfranchized everywhere.
But being at home everywhere
Is to be at home nowhere.

The Son of Man hath not where
To lay his head.
Even the dead
Have laid their heads
Where they're not.

We've had far to go
And have further.
There's no emptiness
Like where home was,
And it won't get less.

In the tale the prodigal son
Staggers back home
And sees his father running.
We've gone further—
We're not who we were.

Our father's stranger too, no longer
In that Old Folks' Home.
We might not recognize him—
Or, shocked, try to revise him,
Perhaps disguise him?

To meet an old friend
After forty years
Is to know, though,
That what you've lost in going
Doesn't go.

LI

People that swim clambered
Onto the land.
Fins elevated into wings
And became the status quo
For excitable birds.
The tiny ampersand
That was crouching on to man
Had far to go.

My patriarch was a man
Called Australopithecus.
He bounced inside his gland
The sperm that quickens us.
And did he have a soul?
Or when the angels scanned
The daughters of men
And saw that they were beautiful
Did something non-indigenous
Scintillate inside that skull?

When One had made it all he saw
It was beautiful and true and good.
Adam's Law
Has split us into two.
And was the tree
That fruited on the spine
Of neolithic or Cro-Magnon man
A breaking awareness of a larger brain,
Or was it law
To help the ploughman to retain
The bread his sweating brow
Had toiled to win?

Round good and evil
Winds the Devil
Our Accuser
With gifts of clothes

And the *felix culpa*,
Trading lust for love,
With loss of Eden, all our woe,
And Cain's
War for grain.

Since Satan's a lawyer
And a gentleman judge
How can he tolerate
That the first Adam,
A stone-pitching primate,
Can be the second
In a Paradise
Of forgiven vice?

Through the gate
The Scandalous Word
Conducts each malefactor:
Past the flaming sword
And the elder brother
With his unprodigal grudge.
Unfair—
The love that makes
The lost and last the heir:
Last here, first there,
Inheriting without merit
An Eden of brotherhood
And incompetence for good.

Beyond the scenery
The in–itself
And its particular machinery
Still elude us. But
Whatever everything is
It still includes us,
Eyeless creatures prodding
At an elephant's toe,
Or a foetus in the amniotic fluid,
Listening to a mother's radio.

Mother science is
The prodigal one
The Father welcomes in
With healing and revealing
Learned studying among husks,
Finding in the musk of swine,
Especially there,
The absconded divine.

If moments occur
When eyelids seem to open,
Eyes even then
Have to learn vision.
These overarching arches are
The womb of Mary.
We look towards the hymen,
The eastern light
Stained blue like sky
Or blue like Mary
Against the particular night.

LII

Invisible reader, impossible God,
There are times we dream you back to being
An irascible blustering Yahveh, haranguing
A gang of bedouin, but not too often.
I try to listen to your presence
Quietly as an astronomer, an absence in the room,
The mother and father I imagined in the dark
When I was sick, who never came.
You're the pebble on the beach with spread
Arms and a navel, the fearful passage
Past the graveyard, a thorn in the flesh,
The psychiatrist who buggered off while listening,
Yet in spite of everything somehow managing
To see me too in my deliberate vanishing.

Note

I began to go to Southwark Cathedral in the late sixties. The 'plant', as they say nowadays, was rather attractively run-down: the flagstones for kneeling at the rail had no sharp corners, were worn concave by hundreds of years of worshippers. Spiritually, it was a furnace of South Bank ideas, warming a rather tiny, conservative, and cold congregation. Its canons had a six-week run each in charge of the services and, since they straddled from something suspiciously like spiritualism at one end to something suspiciously like the death of God at the other, with orthodoxy and lectures on Shakespeare in the middle, there was endless stimulation, and delight. As I once said to the Provost, 'You have canons to the right of you and canons to the left of you.' Presiding and occasionally visible were Bishops Mervyn Stockwood, Hugh Montefiore, and John Robinson. Today it is a restored fabric, with gorgeous music, a larger congregation and quieter clergy; I spend fewer of my Sundays in London and am forgotten there; but I still go whenever I can. Many thoughts, observations, memories, inspirations, and visitations go through your heart as you stand there on Sunday after Sunday, and I thought for me it would be a form of prayer to try to 'kiss' some of these 'as they fly'. I hope that *Letters in the Dark* will give a taste of that 'glorious liberty of the children of God' the Anglican Church so tolerantly and patiently nurtures, knowing it takes open doors and a lifetime to bring a person to God.

Though the number of items equals the weeks of the year, I've avoided the temptation to fit the whole poem into an arbitrary plan, such as the Church Year or a tourists' guide—following instead an intuitive plan corresponding to how things happen in the mind at moments during celebration. I've occasionally incorporated almost direct quotations, and even direct ones, into my text, generally allusions to the lives or words of persons in the poem. The two sources most pillaged were Alexander Whyte's introductory material to his edition of Lancelot Andrewes's *Private Devotions* (Oliphant, Anderson and Ferrier, 1896)—the poem on Andrewes was practically constructed out of this—and *Southwark Story* by Florence Higham (Friends of Southwark Cathedral, 1955). The poem on William Austin, a parishioner whose poems, edited by his wife after his death in 1636, have recently been reissued for the first time, is a montage of quotations from Florence Higham's book. The poem on St Cuthbert was adapted from Bede, *De Vita et Miraculis S. Cudberti*, as translated by Helen Waddell in *Beasts and Saints* (Constable, 1934). The curious reader would find details from *Honest to God* by John Robinson (SCM Press, 1963) and *Chanctonbury Ring* by Mervyn Stockwood (Hodder & Stoughton, 1982); also from *The Shaking of the Foundations* by Paul Tillich (Penguin, 1957) and *Selected Letters* by Rainer Maria Rilke, trans. R. F. C. Hull (Macmillan, 1946).